Indiana
Holt Science & Technology: Gr. 6
Test Preparation Workbook

HOLT, RINEHART AND WINSTON
A Harcourt Education Company
Orlando • Austin • New York • San Diego • Toronto • London

STAFF CREDITS

Director of Special Projects: Suzanne Thompson

Managing Editor: Jenvieve Eyre

Senior Editor: Tressa Sanders

Editor: Jennifer Schwan

Printed in the United States of America

ISBN 0-03-039783-9

1 2 3 4 5 6 7 8 9 032 08 07 06 05 04

Contents

Preparing Students for the Indiana Statewide Testing for Educational Progress

Introduction

This practice book consists of multiple-choice and constructed response questions, representative of the questions on the ISTEP+ Assessment. While this book prepares students for the ISTEP+, no questions contained within are actual ISTEP+ items, nor is this preparation book exactly like the ISTEP+ Assessment.

These practice questions should be used as a part of the local assessment program. A student's performance on the practice questions can serve as an indication of those areas that need additional review and practice. The breadth of content coverage included in this booklet provides teachers an opportunity to assess their students' strengths and weaknesses, as well as the strengths and weaknesses of the overall science program.

Because each page is correlated to a specific chapter or unit of *Indiana Grade 6 Holt Science and Technology,* finding support materials for students to use for review and reinforcement is an easy task. By referring to the relevant section of the textbook, a student can review fundamental science vocabulary, concepts, and principles, and use the questions at the end of the section or chapter for additional review and testing.

Name________________________________ Date____________ Class________________

Chapter 1 Standardized Test Preparation

1. Based on the graph to the right, what conclusion can you make about the human population?

   A. It changes very slowly over time.

   B. The population growth rate has increased in the last 150 years.

   C. Most of the people in the world live in cities.

   D. From 1850 to 2000, the population increased by 4 billion people.

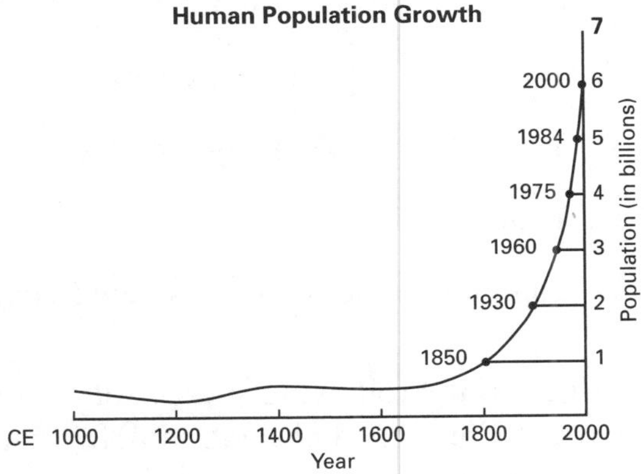

2. Computers are often used in simulations to prepare a

   A. theory.

   B. physical model.

   C. mathematical model.

   D. control variable.

3. A theory is

   A. a possible answer to a question.

   B. a statement of cause and effect that can be used to test a hypothesis.

   C. a unifying explanation for a broad range of hypotheses and observations.

   D. an explanation for the results of an experiment.

4. To test a hypothesis, a scientist must first

   A. communicate results.

   B. identify the variable.

   C. draw conclusions.

   D. analyze results.

Name______________________________ Date__________ Class______________

Chapter 1 Standardized Test Preparation

5. If evidence from an experiment contradicts a hypothesis, then the new data

   A. makes the hypothesis useless.

   B. is used to revise the hypothesis.

   C. is not valid.

   D. cannot be reported until a new hypothesis is developed.

6. Which of the following sources of information would be least likely to be a good place to look for scientific information about space science?

   A. www.nasa.gov

   B. www.startrek.com

   C. www.ucla.edu

   D. www.sciencenews.org

7. Which of the following would NOT employ scientists?

   A. zoo

   B. automobile manufacturer

   C. government agencies

   D. They all employ scientists.

8. Of the places below, where would a scientist be likely to work?

   A. hospital

   B. farm

   C. oil refinery

   D. all of the above

Name________________________ Date_________ Class_____________

Chapter 1 Standardized Test Preparation

Answer the following questions on a separate piece of paper.

1. Explain why models are used to represent scientific concepts and describe some of the limitations of models.

2. Why is it important for scientists to communicate results of a scientific investigation? What kinds of communication do scientists use to publicize their results?

Name________________________ Date__________ Class____________

Chapter 1 Standardized Test Preparation

Answer the following questions on a separate piece of paper.

3. You see an ad that says that "Superpaper" brand of paper towel absorbs 50% more water than other brands. Design an experiment to test this claim.

4. Why has the model of the atom that scientists use changed since 1897? Does the change mean that the old model was wrong and the new model is right?

Name______________________ Date__________ Class__________

Chapter 2 Standardized Test Preparation

1. The tendency of some minerals to break along flat surfaces is called

   A. luster. C. density.

   B. streak. D. cleavage.

2. The chart shown shows the results of a laboratory experiment in which a scientist identified some of the properties and uses of 3 minerals. How did she conduct the experiment that determined the streak of each mineral?

Mineral	Magnetite	Quartz	Sulfur
Hardness	6	7	2
Color	Black	Transparent or in a range of colors	Lemon yellow to yellowish brown
Streak	Black	Colorless	White
Luster	Metallic	Glassy	Greasy
Uses	A source of iron used to make steel	Used in making glass and electronic equipment	Used in fungicides, industrial chemicals, and rubber

   A. She immersed the mineral in water.

   B. She broke the mineral along cleavage lines.

   C. She rubbed the mineral against a hard plate.

   D. She scratched the mineral with a harder mineral.

3. This chart shows Mohs hardness scale for minerals. Students used this chart during a laboratory experiment to determine the hardness of a variety of minerals. Use the hardness scale to determine which of the following minerals is the hardest.

Mineral	Hardness	Mineral	Hardness
Talc	1	Feldspar	6
Gypsum	2	Quartz	7
Calcite	3	Topaz	8
Fluorite	4	Corundum	9
Apatite	5	Diamond	10

   A. fluorite C. calcite

   B. quartz D. feldspar

4. Which of the following properties makes a diamond useful as a cutting tool?

   A. density C. cleavage

   B. hardness D. streak

Name______________________ Date__________ Class______________

Chapter 2 Standardized Test Preparation

5. Which of these stones is too common to be considered a gemstone?

   A. sapphire

   B. aquamarine

   C. quartz

   D. diamond

6. The impacts of mining on the environment can be reduced by all of the following techniques, except

   A. using surface mines instead of subsurface mines.

   B. reusing or recycling metals.

   C. using synthetic diamonds instead of natural diamonds.

   D. reclamation of mined lands.

7. A mineral deposit large enough and pure enough to be mined for a profit is

   A. quartz.

   B. ore.

   C. crystal.

   D. silicate.

8. Which of the following is **NOT** a type of surface mine?

   A. open pit

   B. strip mine

   C. quarry

   D. pluton

Name________________________ Date __________ Class ____________

Chapter 3 Standardized Test Preparation

1. When water, ice, wind and heat break down rock and shape the surface of the Earth, the process is called

   A. weathering.
   B. erosion.
   C. deposition.
   D. uplift.

2. Air chemically weathers rock because

   A. the air molecules are abrasive.
   B. air fills cracks in the rock and later expands, causing the cracks to enlarge.
   C. the oxygen in the air combines with elements in the rock in a process called oxidation.
   D. the rock cracks after absorbing too much oxygen in a process called oxidation.

3. In farming, the purpose of a cover crop is to

   A. prevent soil erosion and replace soil nutrients.
   B. speed up the process of mechanical weathering.
   C. hide other crops from view.
   D. generate extra income for the farmer.

4. Look at the drawing shown. The water falling to Earth in this drawing is probably

   A. made of ozone.
   B. oxygenated.
   C. purified.
   D. acid precipitation.

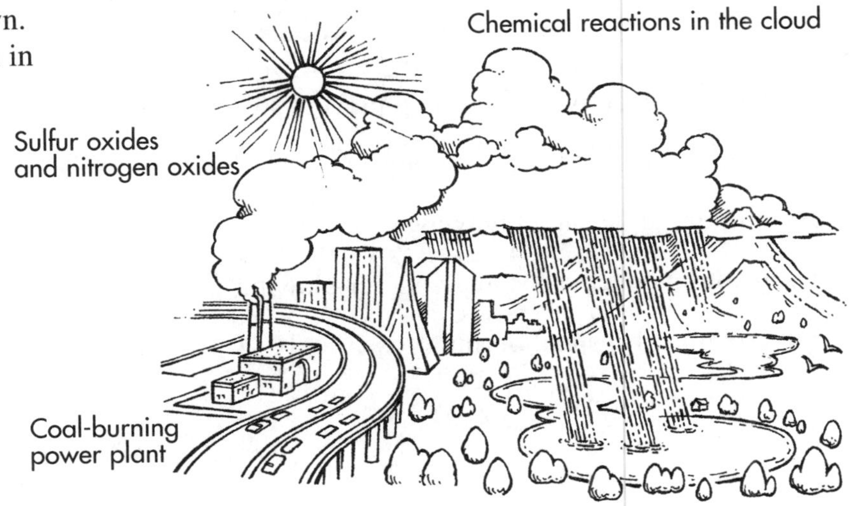

Name________________________ Date __________ Class ____________

Chapter 3 Standardized Test Preparation

5. Poor soil conservation practices will

   A. not affect farmers, as topsoil is an inexhaustible resource.

   B. exhaust Earth's limited supply of topsoil.

   C. result in the expensive process of manufacturing new replacement topsoil.

   D. result in excessive oxidation of the soil's humus content.

6. Soils in temperate regions are generally thick and fertile because

   A. there is no mechanical weathering.

   B. there is a large amount of rain.

   C. there is no chemical weathering.

   D. there is an abundance of chemical and mechanical weathering.

7. A scientist compiled data from numerous field investigations to create the pie chart showing the various causes of soil damage. What percentage of soil damage is caused by wind and water erosion?

   A. 16%

   B. 28%

   C. 56%

   D. 84%

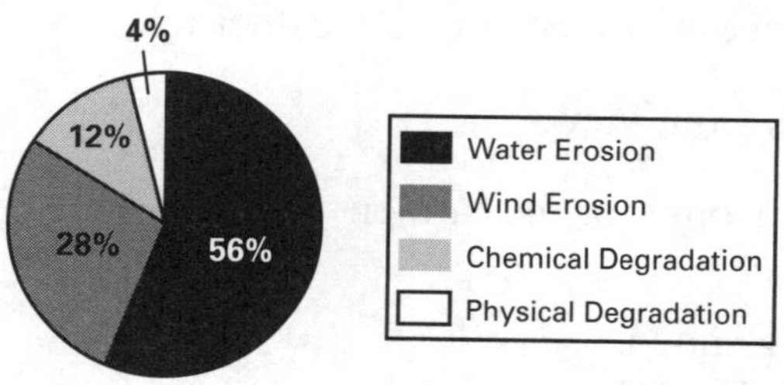

8. Which human activities can help prevent soil erosion?

   A. strip mining and deforestation

   B. building cities and highways

   C. planting cover crops and contour plowing

   D. growing crops and burning fossil fuel

Name ________________ Date ________ Class ________

Chapter 4 Standardized Test Preparation

1. A watershed is a region of land
   - A. that is drained by a river system.
   - B. where rainwater accumulates.
   - C. where the water table extends above Earth's surface.
   - D. where the river water from tributaries accumulates.

2. A rejuvenated river is formed where
   - A. the land is raised by the buildup of ground-water pressure.
   - B. tectonic activity raises the land.
   - C. the land sinks due to gravitational forces.
   - D. new tributaries feed into an old river.

3. A scientist prepared this graph showing the discharge of the Yakima River in the state of Washington as part of a field investigation. The Yakima River flows eastward from the Cascade Mountains to the Columbia River. What is the lowest mean daily discharge recorded during the period of time covered by the graph?

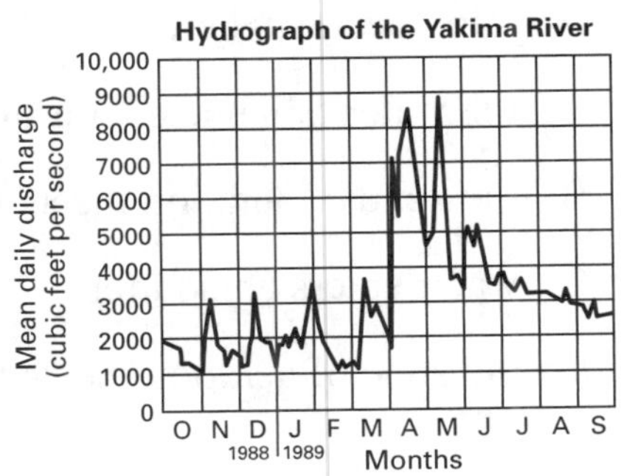

   - A. about 1000 ft^3/s
   - B. about 3200 ft^3/s
   - C. about 770 ft^3/s
   - D. about 8800 ft^3/s

4. The graph above indicates that the discharge of the Yakima River peaks during the months of April and May. Analyze the explanations below and determine which is the most reasonable explanation for the peak in discharge rate.
   - A. Tectonic activity forces ground water to the surface during the months of April and May.
   - B. Runoff from melting winter snows in the Cascade Mountains occurs during the warmer spring months.
   - C. A large winter storm deposited a lot of rain into the watershed.
   - D. The device taking the flow measurements was broken during the months of April and May.

Name______________________________ Date__________ Class____________

Chapter 4 Standardized Test Preparation

5. Which of the following is **NOT** an example of deposition?

   A. the formation of an alluvial fan

   B. a river that overflows its banks and spills out onto a flood plain

   C. the formation of a delta

   D. a river that cuts a deep channel in limestone

6. What is an aquifer?

   A. a machine that purifies ground water

   B. a rock layer that stores ground water

   C. ground water that flows through an alluvial fan

   D. water that remains trapped in caves

7. A spring occurs where

   A. the water table meets Earth's surface.

   B. the zone of aeration meets Earth's surface.

   C. water enters an aquifer.

   D. the zone of aeration meets the zone of saturation.

8. Lakes form in areas where

   A. wells are drilled.

   B. the water table is below the Earth's surface.

   C. the water table is above than the Earth's surface.

   D. there are placer deposits.

Name________________________ Date_________ Class_____________

Chapter 5 Standardized Test Preparation

1. What is the initial energy source for ocean currents and wind?

   A. the rotation of the Earth

   B. the moon's gravity

   C. the sun

   D. continental deflections

2. The sun's energy causes a large amount of ocean water to evaporate. This increases the salinity of the water at the ocean's surface, causing the water to become denser and sink to the ocean floor. This will most likely trigger

   A. a surface current.

   B. El Niño.

   C. a spring tide.

   D. a deep current.

3. Although the British Isles are at a high latitude, the climate in the isles is very mild. What is the most likely reason for this?

   A. The British Isles receive a great deal of sunshine.

   B. The British Isles are located near a warm-weather surface current.

   C. The British Isles experience a very short winter.

   D. The British Isles are located near a cold-water surface current.

4. Southern California has cooler temperatures than Arizona, which is at the same latitude but farther inland. What causes this temperature difference?

   A. a cold-weather surface current off of California's coast

   B. winds formed high in the California mountains

   C. the shade provided by the large number of trees in California

   D. the large longshore current experienced by California

Name_________________________ Date__________ Class_____________

Chapter 5 Standardized Test Preparation

5. As part of a field investigation at the nearby Ocean Science Institute, Steven analyzed a sample of ocean water. The pie chart summarizes the relative amounts of dissolved salts found in the sample. What percentage of the dissolved salts are substances other than chlorine and sodium?

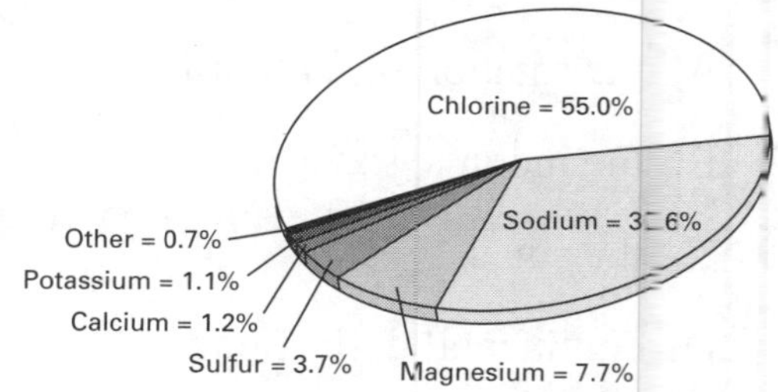

A. 0.7% C. 13.7%

B. 14.4% D. 85.6%

6. Based on Steven's pie chart, which of the following is **NOT** a valid conclusion?

   A. Because there is a greater amount of chlorine in the sample than sodium, there must be additional sources of chlorine in the environment besides the compound sodium chloride.

   B. The dissolved sodium chloride found in the sample is a different compound than the sodium chloride used in cooking.

   C. Magnesium salts are more than twice as common as sulfur salts.

   D. The ocean tastes salty because the most abundant dissolved solid is sodium chloride.

Characteristics of Several Ocean Zones

Zones	Description
Intertidal	air, sun, and water exposure; crashing waves
Neritic	water depth less than 200 m; lots of sunlight; relatively warm water
Benthic	very deep water; no light; cold except near thermal vents that emit heat and chemicals

7. Constance made the table above during a field investigation on ocean life. Which zone would an organism most likely be found in if it has no eyes and can live in water as hot as 80°C?

   A. intertidal zone C. benthic zone

   B. neritic zone D. intertidal or neritic zones

Name______________________ Date__________ Class______________

Chapter 5 Standardized Test Preparation

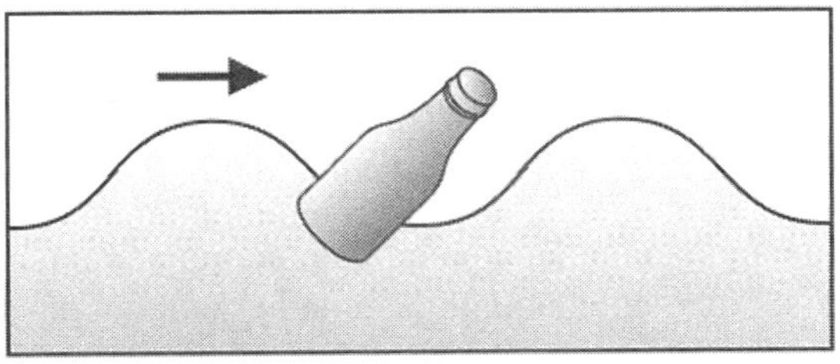

8. Look at the figure above. It shows a bottle George threw into the water to explore the effect of waves on floating objects. As he watches the interaction of the waves and the bottle, which of the following is the most valid conclusion about what is moving across the ocean?

   A. The waves are moving across the ocean.

   B. The bottle is moving across the ocean.

   C. The water is moving across the ocean.

   D. The bottle and water are moving together across the ocean.

9. Based on the figure above, what can be inferred about the direction the wind is blowing?

   A. The wind is blowing from right to left.

   B. The wind is blowing from left to right.

   C. The wind is blowing from top to bottom.

   D. Wind is not responsible for the direction of waves.

10. What type of tide is produced when the Earth is directly between the sun and moon?

   A. neap tide C. low tide

   B. high tide D. spring tide

11. If the moon is full tonight, when will the next neap tide occur?

   A. tomorrow C. in 2 weeks

   B. in 1 week D. in 28 days

Name ______________________ Date __________ Class ____________

Chapter 6 Standardized Test Preparation

1. Which layer of the atmosphere contains most of the atmosphere's total mass?

   A. troposphere C. mesosphere

   B. stratosphere D. thermosphere

2. The ozone layer

   A. blocks 100% of the sun's visible radiation

   B. is made up of nitrogen compounds

   C. absorbs damaging ultraviolet radiation from the sun

   D. is made up of the same form of oxygen that humans breathe

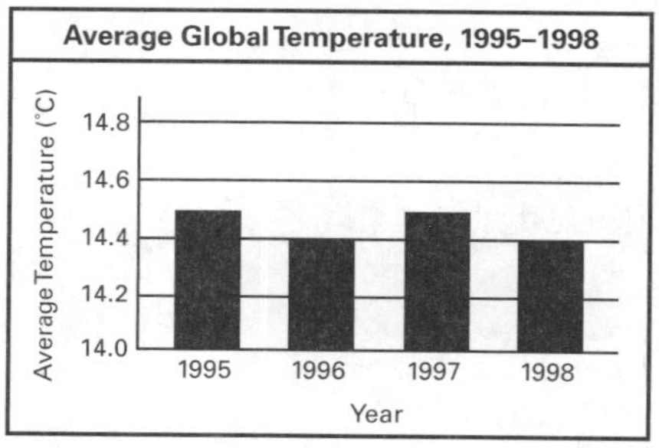

3. Climatologists assembled this graph of the change in average global temperature from data collected over a 4-year period. What can you conclude about average global temperatures?

   A. Over time, average global temperatures are increasing.

   B. Over time, average global temperatures are decreasing.

   C. Over time, average global temperatures are decreasing.

   D. Over time, average global temperatures are staying about the same.

4. Air pollution from cars and factories has a harmful effect on

   A. trees

   B. fish

   C. polar bears

   D. all of the above

Name____________________________ Date__________ Class______________

Chapter 6 Standardized Test Preparation

5. The two main types of winds are

   A. high pressure and low pressure.
   B. northerly and southerly.
   C. easterly and westerly.
   D. local and global.

6. The jet stream

   A. is a narrow belt of high-speed winds in the lower troposphere.
   B. affects the movement of some storm systems on Earth.
   C. always blows from the poles toward the equator.
   D. blows in areas known as the horse latitudes.

7. What term is used to describe the pollutants that are put directly into the air by human or natural activity?

   A. primary pollutants
   B. secondary pollutants
   C. tertiary pollutants
   D. acid precipitation

U.S. Carbon Dioxide Emissions* from Fossil Fuel Energy Consumption

Energy Sector	1997	1998	1999
Residential	289	289	290
Commercial	241	244	244
Industrial	490	480	481
Transportation	474	482	496
Total	1493	1495	1511

*Emissions are given in millions of metric tons of carbon.
Source: Energy Information Administration

8. Based on the information in the table, which of the following is **NOT** a valid conclusion about carbon dioxide emissions?

   A. U.S. carbon dioxide emissions increased each year from 1997 to 1999.
   B. The commercial energy sector is the smallest source of carbon dioxide emissions.
   C. The transportation and industrial energy sectors are the largest sources of carbon dioxide emissions.
   D. Total carbon dioxide emissions have always exceeded 1500 million metric tons of carbon.

Name_______________________ Date__________ Class___________

Chapter 7 Standardized Test Preparation

1. In the illustration of the water cycle at right, at what point does water vapor gain energy?

   A. evaporation

   B. condensation

   C. precipitation

   D. water vapor never gains energy.

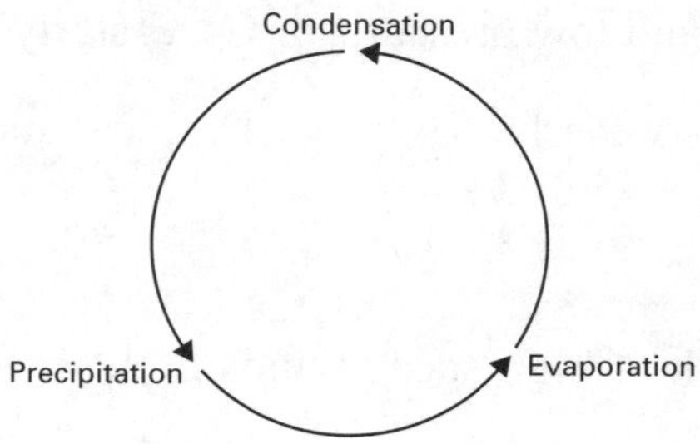

Day	Relative Humidity	Temperature (°C)
1	55%	23°
2	96%	32°
3	83%	30°
4	71%	26°

2. Jason decided to conduct a field experiment on humidity. He compiled his data in the table above, which shows the predicted relative humidity and temperature for 4 days. Based on the information provided in the table, predict on which day the most condensation will form on Jason's window.

   A. Day 1 C. Day 3

   B. Day 2 D. Day 4

3. Using the information he has collected during his field investigation, Jason can make what conclusion about the relationship between temperature and humidity?

   A. The amount of water air can hold increases as temperature increases.

   B. The amount of water air can hold decreases as temperature increases.

   C. Air cannot hold water at any temperature.

   D. Temperature and humidity are unrelated.

Name ______________________ Date __________ Class ______________

Chapter 7 Standardized Test Preparation

4. Titus was doing field research for a school report on meteorology. He studied the sky above his house over a period of two weeks in an effort to find out about different kinds of clouds, how they interact, and what might have caused them to form. As he compiled the results of his research, he created this simple diagram. Which cloud type did Titus represent with the letter A?

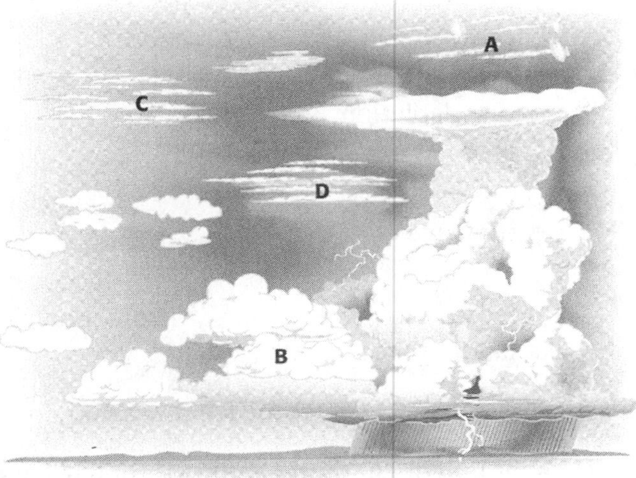

   A. stratus

   B. cumulus

   C. nimbus

   D. cirrus

5. Titus labeled C and D separately in his diagram even though they are both

   A. nimbus clouds.

   B. cumulus clouds.

   C. stratus clouds.

   D. cirrus clouds.

6. What is a front?

   A. the top of a cloud

   B. the top of a weather map

   C. the boundary between air masses

   D. a large body of warm air

7. Which of the following conditions contributes to the decrease in a hurricane's strength as it moves from the ocean onto land?

   A. the lack of warm, moist air over land

   B. the uneven land surface

   C. rising hot air from the land

   D. sinking hot air from the land

Name______________________________ Date__________ Class______________

Chapter 7 Standardized Test Preparation

8. Which of these hazards can be prevented by technological solutions?

   A. tornadoes

   B. ozone depletion

   C. hurricanes

   D. earthquakes

9. The best way to protect yourself from a hurricane that is predicted to make landfall near you is to

   A. nail plywood over your windows and doors.

   B. stay indoors until the entire storm has passed.

   C. travel to a shelter located away from the coast.

   D. make sure that you have flashlights and blankets.

10. During a tornado warning, you should

    A. evacuate to higher ground.

    B. go into a basement or windowless room.

    C. go outside and lie down in a deep ditch.

    D. drive away from the area as quickly as possible.

11. Weather forecasts are predictions based on

    A. data about what happened in previous conditions in the past.

    B. a meteorologist's experience and training.

    C. measurements of temperature, wind, and precipitation.

    D. all of the above.

Name ______________________ Date __________ Class ______________

Chapter 8 Standardized Test Preparation

1. Ancient Chinese cultures could predict

   A. meteor showers.

   B. weather patterns.

   C. eclipses.

   D. planetary motion based on seasonal changes.

2. Galileo Galilei was the first person to

   A. write a book about astronomy.

   B. notice gravitational pull.

   C. use a telescope.

   D. create a modern calendar.

3. Who explained that gravity keeps objects in orbit?

   A. Ptolemy

   B. Copernicus

   C. Brahe

   D. Newton

4. In the 1920s, Edwin Hubble discovered that

   A. the planets stay in orbit because of gravity.

   B. the Earth orbits the sun.

   C. a telescope can be used to view objects in space.

   D. there are galaxies beyond the Milky Way.

Name______________________________ Date__________ Class______________

Chapter 8 Standardized Test Preparation

5. What causes the constellations that are visible in the night sky to change as the seasons change?

   A. Earth revolving around the sun

   B. the light from the stars bending as it travels through Earth's atmosphere

   C. Earth rotating on its axis

   D. the expansion of the universe

6. A light year is

   A. 365 days.

   B. the distance from Earth to the next star.

   C. the amount of light a star produces in a year.

   D. the distance light travels in a year.

7. Sam is planning an investigation of the stars visible from his yard. Which instrument would Sam probably **NOT** select to see the stars?

   A. reflecting telescope

   B. thermometer

   C. pair of binoculars

   D. refracting telescope

8. In non-optical astronomy, we sense infrared radiation as

   A. motion.

   B. heat.

   C. electric shock.

   D. air.

Name ______________________ Date __________ Class ____________

Chapter 9 Standardized Test Preparation

1. Our solar system

   A. is not made up of smaller planetary systems.

   B. includes the planets, but not the sun.

   C. is made up of the inner solar system, the outer solar system, and other smaller systems.

   D. is the only known solar system in the universe.

2. The study of which two planets is the most beneficial in allowing scientists to better understand Earth?

   A. Mars and Jupiter

   B. Mercury and Pluto

   C. Saturn and Venus

   D. Venus and Mars

Data for the Outer Planets

Planet	Diameter (km)	Average Distance From the Sun (km)	Period of Revolution (Earth Years)
Jupiter	142,984	778,600,000	12
Saturn	120,536	1,433,500,000	29
Uranus	51,118	2,872,500,000	84
Neptune	49,528	4,495,100,000	164
Pluto	2,390	5,869,700,000	248

3. Chris made the table above as part of a field investigation on the solar system. According to his data, which outer planet has the largest diameter?

   A. Neptune

   B. Jupiter

   C. Saturn

   D. Uranus

4. Based on Chris's table, which of the following is **NOT** a valid conclusion?

   A. In general, each planet is twice as far from the sun as the next closest planet.

   B. The period of revolution increases as the average distance from the sun increases.

   C. In general, the diameter of the planets decreases as the average distance from the sun increases.

   D. In general, the diameter of the planets decreases as the period of revolution increases.

Name____________________________ Date__________ Class______________

Chapter 9 Standardized Test Preparation

5. Of these planets, the one that would be classified as a terrestrial planet is

   A. Mercury.

   B. Neptune.

   C. Saturn.

   D. Pluto.

6. All of the following factors contributed to the development of life on Earth *except*

   A. its distance from the sun.

   B. the presence of water.

   C. the period of revolution around the sun.

   D. the type of star in the center of the solar system.

7. The planet whose orbit is shaped the least like a circle is

   A. Mercury.

   B. Jupiter.

   C. Neptune.

   D. Pluto.

8. The phases of the moon occur about every 28 days because

   A. the period of rotation of the moon is the same as its period of revolution.

   B. 28 days is approximately the period of revolution around the Earth.

   C. that is the period of a complete cycle of tides.

   D. all of the above

Name______________________________ Date__________ Class______________

Unit 1 Standardized Test Preparation

Answer the following questions on a separate piece of paper.

1. Define erosion and deposition, and explain the changes to the earth's surface resulting from each of the following agents: running water (streams, rivers), glaciers, waves, wind, and mass movement of earth materials.

2. Explain how a Global Positioning System (GPS) works and how it can be used to make more accurate maps. What is a GIS? How do mapmakers use remote sensing, satellite images, and GIS to make predictions about changes in land use and resources?

Name_______________________________ Date __________ Class ____________

Unit 1 Standardized Test Preparation

Answer the following questions on a separate piece of paper.

3. Describe where and how ground water collects. Then explain how different types of landforms or water bodies may form as a result of underground water flows, describing at *least* two examples of each type.

4. What is a planet's period of revolution? Explain the relationship between a planet's distance from the Sun and its period of revolution.

Name______________________________ Date__________ Class______________

Unit 1 Standardized Test Preparation

Answer the following questions on a separate piece of paper.

5. What is a "greenhouse gas" and why are greenhouse gases in the Earth's atmosphere a concern and a controversy? What are some ways to reduce greenhouse gases?

6. Define weather, including the variables that scientists measure in weather observations. Discuss ways in which the water cycle and weather are interrelated.

Name______________________________ Date__________ Class______________

Unit 1 Standardized Test Preparation

Answer the following questions on a separate piece of paper.

7. Explain how two different weather variables affect each other, for example, the effect of temperature on air pressure or precipitation; the effect of temperature on relative humidity; the effect of humidity on precipitation; the effect of air pressure gradients on wind velocity; or the effect of humidity and temperature on cloud formation.

8. What causes tides and tidal cycles on Earth? Compare and contrast the influences of the sun, moon, and Earth's rotation on the tides.

Name ______________________ Date __________ Class ____________

Chapter 10 Standardized Test Preparation

1. A measure of the gravitational pull on matter is called

   A. mass.

   B. weight.

   C. density.

   D. volume.

2. Which tool is needed to measure the volume of a liquid?

   A. metric ruler

   B. graduated cylinder

   C. spring scale

   D. balance

3. Which of these measurements describes a mass?

   A. 29.5 kg

   B. 9.2 m

   C. 12.8 L

   D. 21 K

4. If a liquid has a density of 2.5 g/cm^3 and a mass of 2500 g, what volume does it occupy?

   A. 1.0×10^1 cm^3

   B. 1.0×10^2 cm^3

   C. 1.0×10^3 cm^3

   D. 1.0×10^4 cm^3

Name______________________________ Date__________ Class______________

Chapter 10 Standardized Test Preparation

5. The tendency of all objects to resist any change in motion is called

   A. inertia.

   B. mass.

   C. weight.

   D. volume.

6. Any property of matter that can be observed or measured without changing the identity of the matter is a

   A. proportion.

   B. size.

   C. physical property.

   D. chemical property.

7. Which of the following is **NOT** a physical property of matter?

   A. state

   B. density

   C. solubility

   D. flammability

8. To find the density of an object, you divide its mass by its

   A. weight.

   B. volume.

   C. length.

   D. temperature.

Name ______________________ Date __________ Class ______________

Chapter 11 Standardized Test Preparation

1. A pure substance that cannot be separated into simpler substances by physical or chemical means is

   A. a property.

   B. a compound.

   C. a colloid.

   D. an element.

2. Which of the following is **NOT** a characteristic of metals?

   A. shiny

   B. malleable

   C. brittle

   D. ductile

3. Which of the following is **NOT** a compound?

   A. carbon dioxide

   B. water

   C. table salt

   D. hydrogen

4. One way that mixtures differ from compounds is that mixtures

   A. can sometimes be separated by filtration.

   B. are pure substances.

   C. cannot be broken down through physical changes.

   D. are not composed of elements.

Name ______________________ Date __________ Class ______________

Chapter 11 Standardized Test Preparation

5. A combination of two or more substances that are not chemically combined is called a(n)

   A. mixture.

   B. compound.

   C. element.

   D. pure substance.

6. In a saltwater solution, salt is the ____ and water is the ____.

   A. solute; solvent

   B. solvent; solute

   C. solid; liquid

   D. colloid; solvent

7. Early philosophers thought the basic substances of matter included all of the following *except*

   A. earth.

   B. water.

   C. air.

   D. salt.

8. Elementary particles are subatomic particles that

   A. are the components of all matter.

   B. are the basic units of an element.

   C. have clearly defined sizes and structures.

   D. can be seen with electron microscopes.

Name________________________________ Date__________ Class______________

Chapter 11 Standardized Test Preparation

9. Which of the following statements does NOT apply to Antoine Lavoisier?

   A. He is often called the father of modern chemistry.

   B. He was a French scientist who lived in the late eighteenth century.

   C. He help disprove the idea that there is an element called phlogiston.

   D. He discovered a number of elements, including helium.

10. In a mixture of chemicals, all of the components

    A. can be separated by physical methods.

    B. consist of particles that can be seen without magnification.

    C. are chemically combined.

    D. can be separated by a magnet.

11. In the chemical compound carbon dioxide, the ratio of carbon atoms to oxygen atoms is

    A. 1:1.

    B. 2:1.

    C. 1:2.

    D. 2:2.

12. If the density of substance X is twice that of substance Y, then what is the mass of 50 mL of substance Y

    A. one-half the mass of 50 mL of substance X.

    B. one-quarter the mass of 50 mL of substance X.

    C. equal to the mass of 100 mL of substance X.

    D. one-half the mass of 100 mL of substance X.

Name______________________________ Date__________ Class______________

Chapter 12 Standardized Test Preparation

1. The energy associated with the movement of an object is

   A. potential energy.

   B. kinetic energy.

   C. nuclear energy.

   D. magnetic energy.

2. Sam studied kinetic and potential energy by observing apples on a tree during a field investigation. What type of energy do these apples have?

   A. kinetic energy

   B. mechanical energy

   C. gravitational potential energy

   D. sound energy

3. Which of the following is **NOT** an example of potential energy?

   A. a ball traveling at 10 m/s

   B. a compressed spring

   C. stress along a fault line

   D. a stretched rubber band

4. In the picture above, the circles represent molecules. Which of the jars shows the substance with the most thermal energy?

   A. Jar 1

   B. Jar 2

   C. Jar 3

   D. All substances have the same amount of thermal energy.

Name ______________________ Date __________ Class ______________

Chapter 12 Standardized Test Preparation

5. The ability to do work is called

   A. power.

   B. energy.

   C. velocity.

   D. mass.

6. Mechanical energy is

   A. the total energy of motion and position of an object.

   B. the total work done.

   C. the total power available.

   D. velocity × mass.

7. The energy that comes from food is

   A. thermal energy.

   B. chemical energy.

   C. mechanical energy.

   D. electrical energy.

8. A hair dryer converts ____ energy to thermal energy.

   A. kinetic

   B. potential

   C. electrical

   D. work

Name________________________________ Date__________ Class____________

Chapter 13 Standardized Test Preparation

1. Waves that can transfer energy without traveling through a medium are called

   A. sound waves.

   B. electromagnetic waves.

   C. transverse waves.

   D. longitudinal waves.

2. Waves in which the particles vibrate with an up-and-down motion are called

   A. sound waves.

   B. longitudinal waves.

   C. transverse waves.

   D. electromagnetic waves.

3. In a ____ wave, the particles of the medium vibrate back and forth along the path that the wave travels.

   A. longitudinal

   B. transverse

   C. mechanical

   D. light

4. The maximum distance that a wave travels from its rest position is called its

   A. crest.

   B. trough.

   C. amplitude.

   D. wavelength.

Name______________________ Date__________ Class______________

Chapter 13 Standardized Test Preparation

5. The distance between any two adjacent crests or compressions in a series of waves is called

   A. wavelength.

   B. amplitude.

   C. frequency.

   D. speed.

6. The number of waves produced in a given amount of time is the ____ of the wave.

   A. amplitude

   B. wavelength

   C. speed

   D. frequency

7. If a wave has a wavelength of 3 m and a frequency of 5 Hz, its velocity is

   A. 8 m/s.

   B. 5 m/s.

   C. 15 m/s.

   D. 12 m/s.

8. The bending of a wave as it passes at an angle from one medium to another is called

   A. reflection.

   B. refraction.

   C. diffraction.

   D. interference.

Name____________________________ Date__________ Class______________

Unit 2 Standardized Test Preparation

Answer the following questions on a separate piece of paper.

1. If you mix sugar with water, do you create a compound? Give three reasons for your answer. Could you remove the sugar from the water by using a filter? Explain.

2. Explain why a prism causes sunlight to appear as a rainbow.

Name______________________________ Date__________ Class_______________

Unit 2 Standardized Test Preparation

Answer the following questions on a separate piece of paper.

3. Explain how you could change the pitch and frequency of the sound from a guitar.

4. Explain how the wave patterns produced when two waves of water meet are similar to the wave patterns produced when two rays of light interfere with each other.

Name______________________________ Date__________ Class_______________

Unit 2 Standardized Test Preparation

Answer the following questions on a separate piece of paper.

5. Light and sound are both transmitted by waves. Describe how the two kinds of waves are similar and how they are different.

6. Where does electrical energy for your home come from, how does it get there, and what happens when you use electrical energy?

Name ______________________ Date __________ Class ______________

Chapter 14 Standardized Test Preparation

1. In multicellular organisms, cells are arranged into groups that work together to perform a common function. What are these groups called?

   A. systems

   B. bones

   C. tissues

   D. joints

2. From least complex to most complex, which of the following is the correct order of cellular organization?

   A. cell—tissue—organ system—organ

   B. cell—tissue—organ—organ system

   C. organ system—organ—tissue—cell

   D. cell—organ—tissue—organ system

3. In what way do prokaryotes differ from eukaryotes?

   A. Prokaryotic cells lack a cell membrane.

   B. Prokaryotic cells lack a membrane-bound nucleus.

   C. Prokaryotic cells do not carry genetic information.

   D. Eukaryotic cells lack a nucleus.

4. Which of the following structures is **NOT** found in all eukaryotic cells?

   A. cell membrane

   B. vesicles

   C. cell wall

   D. ribosomes

Name_____________________________ Date__________ Class_______________

Chapter 14 Standardized Test Preparation

5. The largest organ in the human body is

   A. the stomach.

   B. the intestines.

   C. the skin.

   D. the heart.

6. Which of these cell component are found in plant cell but not in animal cells?

   A. ribosome and cell membrane

   B. vacuole and mitochondrion

   C. chloroplast and cell wall

   D. Golgi complex and lysosome

7. About how much of the content of a cell is water?

   A. 25%

   B. 50%

   C. 70%

   D. 90%

8. A cell vesicle containing enzymes that destroy damaged cell organelles is called a(n)

   A. endoplasmic reticulum.

   B. mitochondria.

   C. Golgi complex.

   D. lysosome.

Name ____________________________ Date __________ Class ______________

Chapter 15 Standardized Test Preparation

1. Which classification is most specific?

   A. phylum

   B. species

   C. class

   D. genus

2. Which of the following is the correct order of the 8 levels of classification?

   A. kingdom, family, phylum, class, order, genus, species

   B. kingdom, class, phylum, family, order, genus, species

   C. kingdom, phylum, class, order, family, genus, species

   D. kingdom, phylum, order, class, family, genus, species

3. A Siberian Tiger and a common house cat do **NOT** belong to the same

   A. order.

   B. species.

   C. class.

   D. family.

4. Which of the following is **NOT** a kingdom in the classification system currently used by most scientists?

   A. fungi

   B. eubacteria

   C. bacteria

   D. archaebacteria

Name ______________________ Date __________ Class ____________

Chapter 15 Standardized Test Preparation

5. Until the 1700's, scientists classified organisms into only two kingdoms. Today's scientific classification system has six kingdoms. Why have scientists added more kingdoms?

   A. The previous theory of classification did not account for the Latin names of organisms.

   B. Scientists thought there were too many organisms to classify into only two kingdoms.

   C. Scientists found that some living things did not fit into the plant group or the animal group.

   D. Scientists found too many similarities between organisms in different kingdoms.

6. Carolus Linneaus created the scientific classification system in the 1700s and gave organisms Latin scientific names. Which of the following is **NOT** a reason why scientists still use Latin scientific names for organisms?

   A. The Latin scientific name is always the same for a particular kind of organism, no matter how many different common names the organism might have.

   B. Before Linnaeus's classification system, some species had names up to 12 words long.

   C. Latin scientific names allow individuals who speak different languages to discuss the same species without confusion.

   D. Latin scientific names allow different scientists to name an organism differently.

7. An organism that is a eukaryote, but is **NOT** an animal, plant, or fungi belongs to the kingdom

   A. animalia. C. archaebacteria.

   B. fungi. D. protista.

8. Members of which of the following kingdoms can survive where most other organisms cannot?

   A. protista C. archaebacteria

   B. fungi D. eubacteria

Name______________________ Date__________ Class______________

Chapter 16 Standardized Test Preparation

1. When plants make glucose, they are converting the sun's radiant energy

   A. into ATP.

   B. into chemical energy that can be stored.

   C. in a process known as cellular respiration.

   D. into chlorophyll.

2. If a houseplant is placed near a window, the shoot tips will bend toward the direction of the light. What is this response called?

   A. periodism

   B. geotropism

   C. negative tropism

   D. phototropism

3. How is photosynthesis important to consumers?

   A. Consumers use oxygen and glucose produced by photosynthesis.

   B. Consumers supply the light energy needed for photosynthesis.

   C. Consumers supply the water needed for photosynthesis.

   D. Consumers supply the chloroplasts to producers.

4. Fertilization of a flower occurs when

   A. pollen grains land on the stigma.

   B. a sperm cell unites with an egg cell.

   C. the ovary develops into a fruit.

   D. the seeds form.

Name______________________________ Date__________ Class______________

Chapter 16 Standardized Test Preparation

5. Which of the following statements about leaf structure and function is correct?

   A. The cuticle is responsible for delivering water to the cells in the leaf.

   B. Xylem and phloem are responsible for plant reproduction.

   C. Photosynthesis occurs primarily in the vascular tissue.

   D. A stoma is an opening in the outer surface of the leaf through which gases can pass.

6. After fertilization, which structures become seeds?

   A. ovules

   B. eggs

   C. pistils

   D. stamens

7. Which color is **NOT** absorbed by chlorophyll?

   A. red

   B. green

   C. yellow

   D. violet

8. Which of the following compounds is **NOT** a product of photosynthesis?

   A. glucose

   B. carbon dioxide

   C. water

   D. oxygen

Name__________________________ Date__________ Class ____________

Chapter 17 Standardized Test Preparation.

1. When your body becomes infected with a virus, the lymph nodes sometimes become swollen because

   A. they are making more white blood cells to fight the infection.

   B. the virus is attacking them.

   C. more red blood cells are collecting there.

   D. more fluids from the bloodstream are collecting there.

2. Which of the following is **NOT** part of the lymphatic system?

   A. lymph nodes

   B. tonsils

   C. spleen

   D. blood capillaries

3. Two or more tissues working together form a(n)

   A. organ.

   B. system.

   C. cell.

   D. homeostasis.

4. Which of these is not an organ system?

   A. skeletal

   B. endocrine

   C. nervous

   D. lungs

Name_______________________ Date__________ Class______________

Chapter 17 Standardized Test Preparation

5. The correct path of food through the digestive tract is

   A. mouth, esophagus, stomach, liver, small intestine, rectum, anus.

   B. mouth, esophagus, stomach, small intestine, rectum, anus.

   C. mouth, esophagus, stomach, gallbladder, small intestine, rectum, anus.

   D. mouth, stomach, esophagus, small intestine, rectum, anus.

6. Which of the following is **NOT** a function of the liver?

   A. stores nutrients

   B. breaks down toxic substances in the blood

   C. makes bile to break down fats

   D. makes bicarbonate that neutralizes acids

7. The correct path of urine through the excretory system is

   A. kidneys, urethra, urinary bladder, ureter.

   B. kidneys, ureter, urinary bladder, urethra.

   C. urinary bladder, ureter, kidneys, urethra.

   D. ureter, kidneys, urinary bladder, urethra.

8. Blood can be transferred from one person to another as long as which of these are compatible?

   A. plasma

   B. ABO type

   C. white blood cells

   D. platelets

Name________________________ Date__________ Class______________

Chapter 17 Standardized Test Preparation

9. The organ that stores nutrients that are not needed immediately is the

   A. pancreas.

   B. gall bladder.

   C. lower intestine.

   D. liver.

10. The organs involved in the blood-glucose feedback loop are

    A. pancreas and liver.

    B. gall bladder and liver.

    C. thyroid and pancreas.

    D. heart and lungs.

11. The endocrine glands are a group of organs that produce

    A. bile.

    B. ATP.

    C. DNA.

    D. hormones.

12. How many major organ systems exist in your body?

    A. 4

    B. 8

    C. 12

    D. 16

Name______________________________ Date__________ Class______________

Chapter 18 Standardized Test Preparation

1. A fetus develops in a woman's reproductive system, but it is nourished by her

   A. lymphatic system.
   C. immune system.
   B. circulatory system.
   D. endocrine system.

2. The correct path of an unfertilized egg through the female reproductive system is

   A. ovary, fallopian tube, uterus, vagina.
   B. fallopian tube, ovary, uterus, vagina.
   C. ovary, uterus, fallopian tube, vagina.
   D. fallopian tube, uterus, ovary, vagina.

3. Oxygen passes from the mother to the fetus through

   A. the placenta and the umbilical cord.
   B. the amnion.
   C. the vagina.
   D. the uterus.

4. Male and female sex organs become mature

   A. during infancy.
   B. during childhood.
   C. during adulthood.
   D. during puberty.

Name______________________________ Date__________ Class______________

Chapter 18 Standardized Test Preparation

5. A fertilized egg attached to a fallopian tube is called

   A. infertility.

   B. an implantation.

   C. an ectopic pregnancy.

   D. a miscarriage.

6. When a human sperm cell and an egg cell unite, a new kind of cell forms, called

   A. a zygote.

   B. an embryo.

   C. a fetus.

   D. a spore.

7. The length of a normal human pregnancy is about

   A. 18 weeks.

   B. 30 weeks.

   C. 40 weeks.

   D. 50 weeks.

8. Which of these functions does **NOT** occur in a normally functioning placenta?

   A. Wastes are transferred from the fetus to the mother.

   B. Oxygen is transferred from the mother to the fetus.

   C. Nutrients are transferred from the mother to the fetus.

   D. Blood is transferred from the mother to the fetus.

Name______________________________ Date__________ Class______________

Chapter 19 Standardized Test Preparation

1. Which of the following is a noninfectious disease?

   A. strep throat

   B. dysentery

   C. rabies

   D. asthma

2. Which of the following is **NOT** a way you might come in contact with a pathogen?

   A. eating contaminated foods

   B. breathing contaminated air

   C. handling contaminated objects

   D. eating a high-fat diet

3. A substance that helps your body develop immunity to a disease is called

   A. a vaccine.

   B. an antibiotic.

   C. an antipathogen.

   D. a germicide.

4. Which of the following statements about the immune system is **NOT** true?

   A. B cells make antibodies.

   B. T cells coordinate the immune response.

   C. Macrophages eat viruses.

   D. Antigens attach to bacteria and mark them for attack by immune cells.

Name ____________________ Date __________ Class ____________

Chapter 19 Standardized Test Preparation

5. A fever can be helpful in fighting an infection because it
   A. increases the rate of pathogen reproduction.
   B. increases the rate of T cell reproduction.
   C. causes memory cells to form.
   D. causes the blood to flow faster.

6. Your immune system is controlled by
   A. your brain.
   B. your pituitary gland.
   C. T cells and B cells.
   D. a number of cells, tissues and organs throughout your body.

7. Antibiotics are an effective treatment against which of these illnesses?
   A. strep throat
   B. flu
   C. a cold
   D. allergies

8. If a doctor prescribes antibiotics to treat an infection, how long should you take them?
   A. three days
   B. until all your symptoms are completely gone
   C. two weeks
   D. until the entire prescription is used

Name________________________ Date__________ Class______________

Chapter 20 Standardized Test Preparation

1. Which of the following is part of a salt marsh community?

   A. a river that empties into the marsh

   B. the soil

   C. the salt water that washes into the marsh

   D. hermit crabs

2. Which of the following is the correct order of the levels of environmental organization?

   A. organism, population, community, ecosystem, biosphere

   B. organism, community, population, ecosystem, biosphere

   C. organism, community, ecosystem, population, biosphere

   D. organism, population, ecosystem, community, biosphere

3. At each higher level of the food pyramid

   A. the amount of energy decreases.

   B. the number of organisms increases.

   C. the amount of energy increases.

   D. the size of organisms increases.

4. Which of the following organisms is a producer?

   A. mouse

   B. owl

   C. mushroom

   D. grass

Name______________________________ Date__________ Class_______________

Chapter 20 Standardized Test Preparation

5. Which of the following would **NOT** be a limiting factor for an antelope in the Savanna?

   A. water

   B. space

   C. food

   D. size of the antelope population

6. Which of the following is **NOT** an example of a prey adaptation?

   A. camouflage

   B. staying in herds

   C. advertising their poison with bright colors

   D. sharp claws and teeth

7. An example of mutualism is

   A. a remora hitching a ride on a shark.

   B. a tapeworm inside a human's intestine.

   C. beneficial bacteria inside a human's intestine.

   D. a female wasp laying eggs on a tomato hornworm.

8. Which of the following is **NOT** an example of symbiosis?

   A. parasitism

   B. mutualism

   C. commensalism

   D. predation

Name_______________________ Date__________ Class______________

Chapter 21 Standardized Test Preparation

1. The most biologically diverse biome on Earth is

   A. a temperate deciduous forest.

   B. a tropical rainforest.

   C. a savanna.

   D. a coral reef.

2. Why are savannas well suited to support large herbivores?

   A. They have very little biodiversity.

   B. They are areas where grasses are the main plants.

   C. They are dry and hot.

   D. They are cold and dry.

3. Some plants have stems that can hold water and carry on photosynthesis, and almost no leaves. In which type of habitat would this be important?

   A. grassland

   B. rainforest

   C. swamp

   D. desert

4. Which of the following is **NOT** a major abiotic factor influencing the kind of marine biome that exists in an area?

   A. temperature

   B. amount of sunlight penetrating the water

   C. distance from land

   D. type of sand on the ocean bottom

Name____________________________ Date__________ Class______________

Chapter 21 Standardized Test Preparation

5. The most important producers in the ocean are

   A. phytoplankton.

   B. seaweed.

   C. zooplankton.

   D. sea grasses.

6. Coral reefs exist in which of the following ocean zones?

   A. intertidal zone

   B. neritic zone

   C. oceanic zone

   D. benthic zone

7. How are savannas different from temperate grasslands?

   A. The vegetation is mainly grasses.

   B. There is a warm season and a cold season.

   C. There is a wet season and a dry season.

   D. The average yearly temperature is lower.

8. An area where fresh water from streams and rivers spills into the ocean is called

   A. an intertidal area.

   B. an estuary.

   C. a marsh.

   D. a wetland.

Name______________________________ Date__________ Class______________

Unit 3 Standardized Test Preparation

Answer the following questions on a separate piece of paper.

1. Explain the meaning of the term "good nutrition," including how to eat the right amounts of each type of food.

2. Describe at *least* three ways that pathogens can spread from person to person and at *least* three technologies or actions that help to control the spread of pathogens.

Name______________________ Date__________ Class______________

Unit 3 Standardized Test Preparation

Answer the following questions on a separate piece of paper.

3. Identify and describe the purpose of at *least* six major human organ systems.

4. Explain the roles of at *least* five of the following organs in digestion, including how both mechanical and chemical digestion play a role in the digestive process: teeth, salivary glands, mouth, throat, esophagus, stomach, liver, gallbladder, pancreas, small intestine, large intestine, villi. Describe how the digestive system interacts with the circulatory system.

Name______________________________ Date__________ Class______________

Unit 3 Standardized Test Preparation

Answer the following questions on a separate piece of paper.

5. Explain how the body's nervous system uses feedback to regulate body temperature.

6. Describe the benefits that a multicellular structure gives to an organism compared to single cell organisms.

Name______________________________ Date____________ Class_______________

Unit 3 Standardized Test Preparation

Answer the following questions on a separate piece of paper.

7. Identify at *least* five ways in which all cells are alike.

8. Complex organisms have many levels of cell organization. Define each of the following levels: cells, tissues, organs, organ systems, organisms. Provide at *least* one plant and one animal example for each level, and describe how each level is related to the other levels.

Name______________________________ Date__________ Class______________

Unit 3 Standardized Test Preparation

Answer the following question on a separate piece of paper.

9. Explain at least two methods by which flowering plants can reproduce.